The Electrical Utility In The Digital Age

&

Revival of the Science of Electricity in the Digital Age

The Electrical Utility In The Digital Age

&

Revival of the Science of Electricity in the Digital Age

By

Eric P. Dollard

Published by
Emediapress.com
Spokane, Washington

Cover Layout: Aaron Murakami
Transcribed by: Jeff Moe
Edited by: Simon Davies (www.teslascientific.com)

Digital Version 1.0 Released: Oct 26, 2018 & March 26, 2019
Print Version 1.0 Released: May 2021

ISBN-13: 9798746918185

Digital Edition Published by: Emediapress.com
PO Box 10029
Spokane, WA 99209
https://emediapress.com

EPD Laboratories, Inc.
PO Box 10029
Spokane, WA 99209

info@epdlabs.org
https://ericpdollard.com

Support EPD Laboratories, Inc. 501(c)3 non-profit organization with tax-deductible donations at https://ericpdollard.com

Table Of Contents

The Electrical Utility In The Digital Age

Engineering Report
NVE-1

Introduction

Presently, much discussion exists regarding the proliferation of harmonics in the electric utility system. Moreover, the susceptibility of this system to damage or destruction from a well-positioned nuclear E.M.P. is a subject of great concern.

Such problems are not necessarily intrinsic to the process of electric transmission and distribution, but only present themselves when certain configurations exist within the electric utility system. Paramount in any transmission or distribution system is that such system must be a closed system, that is, the electromagnetic boundary condition must be maintained to the highest practical degree. This assures that the entry or exit of extraneous electric forces is minimized.

Presently, with regard to the electric utility system, the electromagnetic boundary condition is being violated in two ways:

1) By the ubiquitous employment of Wye to Wye connected transmission and distribution transformers, and related load configuration

2) Through the extension of a common neutral connection throughout the entire electric utility system

The widespread and enforced adoption of these practices has rendered the electric utility system an open system, or in other words, an antenna system, one of unprecedented proportions. The consequences must be obvious.

It is the objective of the following series of engineering reports to re-establish a fundamental understanding of polyphase alternating current, and then apply this understanding to the analysis of the situation as it exists today with regard to the electric utility system.

The Engineering Dilemma

Electrical engineering in the digital age has taken a turn into the abyss. The engineering foundation so well established through the efforts of Arthur Kennelly, Charles Steinmetz, and other notables, has been outright abandoned. Engineers have been replaced by software technicians, and accordingly the electric utility system has become a morass of harmonics, parasitic oscillations, and stray currents. This malady is presently being carried out with deliberation, and moreover, being made law.

The Situation, Then And Now

Kennelly once remarked that a properly designed three phase power system contains no significant harmonics. This may have been true in the era in which Kennelly lived, but presently the proliferation of harmonics in the electric utility system is rampant. In the manner of Oliver Heaviside, one may ask; If harmonics are a problem, then why allow them to begin with. The child exclaims, "The Emperor wears no clothes!"

In the early electrical age, the principal customer loads were the incandescent lamp of Thomas Edison, and the induction motor of Nikola Tesla. Neither was a source of significant harmonics. What harmonic content did exist was neutralized by the Delta connection of apparatus windings.

The concurrent development of long distance telephone served as a check upon the harmonic content of the electric utility system. In this era the long lines were of open wire construction and utilized voice frequency transmission, both of which are susceptible to power system interference. This situation led to the establishment of an administrative agency to co-ordinate electromagnetic compatibility between power and communication companies.

The Genesis Of The Harmonic Problem

Two principal factors led to the harmonic situation as it exists today in the electric utility system. One was the gradual adoption of a Wye connection of apparatus windings and load configurations; another was the increasing use of fluorescent lighting as a replacement for incandescent lighting.

The general use of a Wye connected transformer configuration grew out of the need to expand the load capacity of existing distribution systems to meet the ever-increasing demand for electric light and power. Delta to Wye system conversion allowed for a 173% increase in capacity while retaining the original transformers. The only additional requirement in this conversion is the system-wide installation of a fourth conductor, or system neutral. An additional benefit derived from the neutral was that the system could be referenced to

Earth potential without the need for auxiliary apparatus, as was necessary in a Delta connected system.

The dominance of the Wye connected load grew out of the proliferation of large office complexes. The enormity of their lighting systems required a shift from the Edison three wire system to a Wye connected four wire configuration, this in order to provide a good phase balance to the electric utility providing the power for illumination. A significant penalty for this conversion is that while the Edison system suppresses the harmonics, the Wye system reinforces the harmonics.

The large-scale generation of harmonics began when the fluorescent lamp replaced the incandescent lamp in these large office complexes. The fluorescent lamp and its reactance ballast are a prodigious source of troublesome harmonics. Ultimately such a load should be Delta connected in order to suppress the proliferation of these harmonics into the electric utility system. However, such connection requires two pole rather than one pole switching and circuit protection devices, which increases the cost of installation.

In the electric age the primary windings of the distribution transformers supplying electric power to these lighting systems were typically Delta connected. This prevented entry of the harmonics into the electric utility system. The penalty was to be increased transformer heating, this is a consequence of confining the load generated harmonics.

Wye Connection Of Transmission Systems

The widespread adoption of Wye connected electric utility systems can be attributed to the efforts of the rural electrification administration, R.E.A. Here the objective was to significantly reduce the cost of delivering electric power to the more remote areas of the United States. Through the employment of the higher transmission voltages for the purpose of distribution, the "copper" cost was greatly reduced. Because no transformer was required to step down this transmission voltage to a distribution voltage, the cost of a distribution substation was obviated. The penalty for this practice was that the transmission system must also be Wye connected.

Another cost savings measure adopted by the R.E.A. program was the implementation of a multiple grounded neutral throughout the entire system. This neutral was made common to both the high side and low side of the distribution transformer, effectively bypassing the transformer. This practice significantly reduced the insulation requirements on an otherwise high voltage system. The penalty for this practice is the exposure of the customer to primary neutral transients, and difficulty in ground fault protection.

The practices instituted by the R.E.A. program significantly enhanced the ability of the resulting electric utility system to interfere with other services. Inevitably, what saved the power company money cost the phone company money, and endangered the customer.

The Electric Utility In The Electronic Age

As society moved from the electrical age into the electronic age the situation of the electric utility system began to deteriorate. In the electronic age an ever-increasing proportion of the load presented to the electric utility system was to involve the conversion between the alternating current used for light and power and the direct current used for electronic devices. The conversion between A.C. and D.C., rectifiers and inverters, is verily a condition of "forcing a square peg into a round hole". Accordingly, this process engenders a plurality of troublesome transients, particularly those associated with the silicon-controlled rectifier (S.C.R.), which ultimately spelled disaster for the electric utility system.

The harmonic malady was propelled forward by a shift in reasoning, this from the established viewpoint of electric transmission as an electromagnetic boundary condition (Maxwell-Heaviside), to the modernistic viewpoint of electric transmission as an electronic current (Lorentz-Einstein). In the modern viewpoint electricity was no longer propagation in the space enclosed by the transmission conductors, but the notion is that electricity is propagated within the transmission conductors themselves. Accordingly, in this view, the space surrounding the transmission conductors is void of any electrical activity. This mindset has made impossible any understanding of electric transmission and has sent electrical engineering back into its dark ages when electricity was regarded as a material substance.

The Electric Utility In The Digital Age

The electric utility in the digital age has arrived at a situation where all deleterious factors have merged into a common motive force. Nikola Tesla would say "Commercial Enterprise has out-run technical competence".

The vanguard is the ubiquitous eradication of the Delta connection of apparatus windings and load configurations. The entire electric utility system is undergoing a retrograde conversion, from what was an interlocked triple phase system, into a trio of independent singe phase systems. Moreover, the primary and secondary windings of all transmission and distribution transformers are being bonded together at their neutral points. This effort effectively bypasses the transformers, hence creating an extensive metallic pathway for the propagation of harmonics, oscillations, and disruptive transients.

In this digital age the load presented to the electric utility system is of a disruptive discharge nature. All contemporary power conversion methodologies employ the integration of switching transients in the conversion process between the utility system alternating current and the electronic (silicon) device direct current. In general, a true A.C. load is becoming non-existent.

It is the intrinsic nature of such transient waveforms to excite oscillations into the electric field surrounding the various transmission and distribution structures. These in turn intensify the already problematic harmonic waveforms. Moreover, the interaction of these electric oscillations with the semi-conductor devices results in a barrage of spurious frequencies. The plurality of these parasitic oscillations tends to merge into a unified oscillation through the common neutral connection, giving the ability to distribute interference on an unprecedented level. This situation alludes to electronic warfare, where consumer devices of foreign manufacture can be regarded as a "Trojan Horse".

It has become evident to many that the aforementioned condition has created an unprecedented level of radio interference, particularly in the A.M. broadcast band. However, unlike the era of long distance telephone, no administrative agency exists to ensure electromagnetic compatibility in this digital age.

Epilogue

It is important to recognize that the ultimate root of the problem here defined is sociological, not technological. The situation today resembles in many ways that which existed at the onset of the dark ages, when the first theory of relatively announced that the Earth is the center of the universe, a notion brutally enforced by the corporate church.

Accordingly, any attempt toward remediation of the electric utility situation will be met with fierce opposition and nothing will change for the better. Such is the fate of the human species, it is written.[1]

---

[1] History, Theory & Practice Of The Electrical Utility System, Eric P. Dollard, https://emediapress.com/utility

REVIVAL OF THE SCIENCE OF ELECTRICITY IN THE DIGITAL AGE

Prologue

In this Digital Age, the theoretical basis provided for Electrical Engineering has been defiled, reduced to an assemblage of delusive pronouncements enforced by a cabal of academic theoreticians. What gives authoritative force to their dictates is the overtly complex and convoluted mathematics which enshrouds them. This mathematics in turn becomes a "rite of passage" in teaching and replacing reason and experiment. Once the idols become digitized, they become law, and consequently electricity has become a lost science, relegated to the same scrap heap as alchemy.

However, in historical perspective, the theoretical basis for electrical engineering had achieved a very high level of development at the onset of the 20th century, this to meet the demands of a rapidly developing commercial enterprise, most notably that of the electro-magnetic telegraph and later electric light and power. Electrical science progressed remarkably well in this era, and thus became an exact science which engendered well thought out theories. The foundations of this science were derived from the Experimental Researches of Michael Faraday and their Mathematical Development by James Clerk Maxwell. This became known as the "Faraday-Maxwell" Theory of Electricity. J. J. Thomson gives an account of the basis of the theory:

"This method is based on the conception, introduced by Faraday, of tubes of electric force, or rather, of electrostatic induction. Faraday, as is well known, used these tubes as the language in which to express the phenomena of the electric field. Thus, it was by their tendency to contract, and the lateral repulsion which similar tubes exert on each other, that he explained the mechanical forces between electrified bodies, while the influence of the medium on these tubes was on his view indicated by the specific inductive capacity in dielectrics." [1]

Essential to the Faraday-Maxwell concept is the existence of a fundamental "Electric Medium," or aether, which fills all space and permeates all matter. Regardless of the fruitfulness of this idea of the aether, and its tubes of electric induction, it was later forcibly driven from electrical theory and replaced with an idol, the "Electron," and its sacred companion, Einstein-Minkowski Relativity.

In the course of human events there always comes a time when certain forces set out to tear apart what became a well-established and thought out process of reasoning. Such was the objective of the modernistic theoreticians. An excellent remark on this condition in historic events is given by E.T. Whitaker:

"Perhaps nothing in the history of natural philosophy is more amazing than the vicissitudes of the theory of heat. The true hypothesis, after meeting with general acceptance throughout a century, and having been approved by a succession of illustrious men, was deliberately abandoned by their successors in favor of a conception so utterly false, and in some of its developments, grotesque and absurd." [2]

Meanwhile, as Gustave Le Bon remarks:

"The mathematicians were drawing up formulae, the physicists were making experiments, and these experiments fitted less with the formulae. So soon as the equations no longer agreed with the experiments, the equations were rectified by imagining "hidden forces," which completely baffled observation." [3]

The modernistic mathematicians ultimately regained the ground they had lost to Faraday, and electrical science was reduced to meta-physics. The three fundamental commandments established are:

I. The abolition of the aether
II. The Enshrinement of a certain set of so-called "Maxwell's Equations"
III. The idolization of the so-called "electron"

A discussion of these three commandments follows.

The First Commandment - The Abolition Of The Aether

Nikola Tesla once remarked, "The discovery of the aether was as significant (in his era) as was the discovery of fire was for primitive man."

However, at the onset of the electronic age, the concept of the aether was rejected with all the vehemence of religious fanaticism. The justification for this was that no physical model could be devised that could survive the test of experiment. This is no surprise since the aether is not physical and accordingly it is outside the realm of physical science. Many resisted this abolition at its onset at the beginning of the 20th century, one was Oliver Heaviside, to quote:

"As one regards the aether, it is useless to sneer at it this time of day. What substitute for it do we have? Its principle fault is that it is mysterious. That is because we know so little about it. Then we should find out more. That cannot be done by ignoring it. The properties of air, so far as they are known, had to be found out before they became known." [4]

Joseph Larmor states the following:

"It therefore may be held that, in so far as theories of the ultimate connection of different physical agencies are allowed to be legitimate at all, they should develop along the lines of a purely electric aether until critics of such a simple scheme are able to point to a definite group of phenomena that requires the assumption of a new set of properties and that moreover can be reduced to logical order thereby. A charge of incompleteness without indication of a better way, is not effective criticism in question of this kind, because, owning to the imperfections of our perceptions and the limited range of our intellectual operations, finally can never be attained." [5]

Larmor continues with:

"From remote ages, the great question with which, since Newton's time, we have been familiar under the somewhat misleading antithesis of contact vs. distance actions, has engaged speculation, -how is it that portions of matter can interact on each other which seems to have no means of connection between them, can a body act where it is not? If we answer directly in the negative, the spatial limitations of substance are to a large extent removed, and the complication is increased. The simplest solution is involved in a view that has come down from the early period of Greek physical speculation, and forms one of the most striking items in the stock of first principles of knowledge which has been struck out by the genius of that age. In that mode of thought, the ultimate reality is transferred from sensible matter to a uniform medium which is a plenum, the ultimate elements of matter consisting of permanently existing vortices or other singularities of motion and strain located in the primordial medium." [6]

At the onset of the "Age of Enlightenment," this concept was revived from the age of antiquity by Rene Descartes in his absolute rejection of "action at a distance" through an empty or vacuous space devoid of any physical properties. He assumed that force cannot be communicated except by actual pressure or impact. "It is thus erroneous to regard the heavenly bodies as isolated in vacant space; around and between them is an incessant conveyance and transformation of energy. To the vehicle of this activity, the name aether has been given." [7]

Following this line of reasoning is that of Michael Faraday, which is a continuous transmission of physical actions through an electrified aether. This was one of the first and most important, in the stock of first principles, in the theory of electricity. To quote Lord Kelvin:

"During the 56 years which have passed since Faraday first offended physical mathematicians with his curved lines of force, many writers and many thinkers have helped to build up the 19th century School of the plenum, one aether for light, heat, electricity, magnetism…" [8]

"Faraday had strong geometric conceptions which he formulated qualitatively. To the more abstract minds of the European theoreticians, these concepts did not appeal at all." [9]

At the onset of the electronic age, Faraday's concept of the electric lines of force were rejected, despite all the physical evidence to substantiate their existence, and the meta-physical notion of action at a distance again reigned supreme. However, action was now presented through so-called "curved space," to circumvent reality.

The common sense appeal to facts demands the retention of the philosophy of Michael Faraday; that is, this electric aether can be expressed in practical terms as an "electric fluid." This concept held supreme in the pioneering days of the science of electricity (Ben Franklin). J. J. Thomson remarks:

"The influence which the notion and ideas of the fluid theory of electricity have, ever since their introduction, over the science of electricity and magnetism, is a striking illustration of the benefits conferred upon this science by a concrete representation of the symbols which in the mathematical theory of electricity define the state of the electric field. Indeed, the services which the old fluid theories has rendered to electricity by providing a language in which the facts of the science can be clearly and briefly expressed, can hardly be over-rated. A descriptive theory of this kind does more than serve as a vehicle for the clear expression of well-known results, it often renders important services by suggesting the possibility of the existence of new phenomena." [10]

Gustave Le Bon provides a similar viewpoint:

"What is called electricity proceeds solely from phenomena from the so-called displacement of the electric fluid or of its elements." [11]

"Electricity will appear as the connecting link between the world of matter and that of the aether." [12]

"Electricity allowed for the connection of two worlds, the ponderable and the imponderable." [32]

"Such is the current theory. It is probable that things happen in a less simple, perhaps in even a very different manner; but when an explanation fits in fairly well with known facts, it is wise to be satisfied with it." [13]

The Second Commandment - Enshrinement Of The So-Called "Maxwell's Equations"

The Faraday-Maxwell theory of electricity has for quite some time served as the principal foundation for electrical engineering theory.

"Maxwell's brilliant synthesis of all electric and magnetic phenomena as well as their interactions into two simple "field equations" was an achievement of singular grandeur and beauty." [14]

Well, maybe not really all electric and magnetic phenomena, and what about the rest of the equations?

"As a matter of historical fact, Maxwell himself never wrote or saw "Maxwell's Equations." [15]

In reality, he conceived twenty mathematical propositions, four of which ultimately were named "Maxwell's Equations," but these were in reality the result of the combined efforts of Heinrich Hertz in Germany, and Oliver Heaviside in England. As it stands, the physicist prefers Hertz, whereas the engineer prefers Heaviside. However, the fundamental units, dimensions, and terminology from which the science and engineering of electricity derived its basis, are essentially as Maxwell formulated them. This is a rather unfortunate circumstance since these formulations exist in a primitive form, which in turn was "set in stone" by the academics before such formulations could develop into a mature theory of electricity. Also unfortunate is how the work of Herman von Helmholtz was swept under the carpet, which led to the unfortunate omissions in electrical theory.

Hertz gives his comments on this:

"In the researches to which I have hitherto referred, the experiments were interpreted from the standpoint which I took up through studying von Helmholtz's papers. In these papers, Herr von Helmholtz's distinguishes between two forms of electric force, the electro-static and the electromagnetic to which, until the contrary is proven by experiment, two different velocities are attributed. An interpretation of the experiments from the point of view could certainly not be incorrect, but it might be unnecessarily complicated. In a special limiting case, Helmholtz's theory becomes considerably simplified, and its equations, in this case, become the same as those of Maxwell's theory; only one form of force remains, and it is propagated with the velocity of Light." [16]

At this historic juncture, the so-called "Velocity of Light" became enshrined, and the electro-static wave faded from existence. But, Nikola Tesla, who worked from the electro-static viewpoint remarks:

"For more than 18 years I have been reading treatises, reports of scientific transactions, and articles on Hertz-Wave Theory, to keep myself informed, but they always impressed me like works of fiction." [17]

Here, it becomes evident that the so-called Maxwell's Theory is not without significant limitations. For example, Charles Steinmetz makes a rather harsh statement on this matter:

"The Maxwellian Theory of the Transformer describes a device that does not exist in practice, but haunts in textbooks and mathematical treatises on transformers." [18]

Consequently, what has become known as "Maxwell's Equations" is in part, of dubious practical value and serves in actuality as a kind of "religious ritual" rather than a fruitful method of analysis.

"At the time of Maxwell's death, which happened in 1879, before he completed his 49th year, much yet remained to be done both in the investigations with which his name is associated; and the energies of the next generation were largely spent in extending and refining that conception of electrical and optical phenomena whose origin is correctly indicated in the name of Maxwell." [19]

The three principal figures involved in this extension, or adaptation, of the work of clerk Maxwell were:

1) Heinrich Hertz
2) Oliver Heaviside
3) J. J. Thomson

These illustrious individuals can rightly be called the "Followers of Maxwell."

"In the decades following the death of Maxwell, his "theory" would be developed in ways which could scarcely have been anticipated. But although every year added something to the super-structure, the foundations remained much as Maxwell had laid them; the doubtful argument by which he had sought to justify the introduction of displacement currents was still all that was offered in their defense. In 1884 however, the theory was established on a different basis by a pupil of Helmholtz, Heinrich Hertz." [20]

Hertz states, in his book on electric waves, the following:

"And now to be more precise, what is it we call the Faraday-Maxwell Theory? Maxwell has left us as a result of his mature thought a great treatise on electricity and magnetism; it might therefore be said that Maxwell's Theory is the one which propounds that work. But such an answer will scarcely be regarded as satisfactory by all scientific men who have considered the question closely. Many a man has thrown himself with zeal into the study of Maxwell's work, and even when he has stumbled upon unwonted mathematical difficulties, has never the less been compelled to abandon hope of forming for himself an all-together consistent conception of Maxwell's ideas." [21]

Oliver Heaviside, the leading proponent of Maxwell's Electromagnetic Theory, comments on the efforts of Hertz to verify this theory:

"Returning to electromagnetic waves, Maxwell's Inimitable Theory of Dielectric Displacement was for a long time generally regarded as speculation. There was, for many years, an almost complete dearth of interest in the unverified parts of Maxwell's Theory..."

"Still, however, they wanted experimental proof. Three years ago, electromagnetic waves were nowhere. Shortly after, they were everywhere. This was due to a very remarkable and unexpected event, no less than the experimental discovery by

Hertz, of Karlsruhe (Now of Bonn), of the veritable actuality of electromagnetic waves in the aether." [22]

Heaviside continues on Maxwell:

"What is Maxwell's Theory? Or, what should we agree to understand by Maxwell's Theory? The first approximation to the answer is to say; There is Maxwell's book as he wrote it; There is his text, and there are his equations: Together, they make his theory. But when we come to examine it closely, we find that this answer is unsatisfactory. To begin with, it is sufficient to refer to papers by physicists, written say during the twelve years following the first publication of Maxwell's treatise, to see that there may be much difference in opinion as to what his theory is. It may be and has been, differently interpreted by different men, which is a sign that is not set forth in a perfectly clear and unmistakable form. There are many obscurities and some inconsistencies…

It is therefore impossible to adhere strictly to Maxwell's Theory as he gave it to the world, if only on account of its inconvenient form. But it is clearly not admissible to make arbitrary changes in it and still call it his." [23]

Professor J. J. Thomson gives a similar view on the problems in expressing the ideas of Maxwell:

"The descriptive hypothesis, that of displacement in a dielectric, used by Maxwell to illustrate his mathematical theory, seem to have been found by many readers neither so simple nor so easy of comprehension as the old fluid theory; indeed this seems to have been one of the chief reasons why his views did not sooner meet with the general acceptance they have since received. As many students find the conceptions of 'displacement' difficult, I venture to give an alternative method of regarding the process occurring in the electric field, which I have often found useful and which is, from a mathematical point of view, equivalent to Maxwell's s Theory.

This method is based on the conception, introduced by Faraday, of tubes of electric force, or rather, of electro-static induction." [24]

It is evident that three distinct contributions were laid out by these followers of Maxwell's Theory. First was the experimental verification by Hertz. Second was the development of a suitable mathematical representation applicable to engineering formulation by Heaviside, and third, an aether physics representation by Thomson.

The problem introduced however, is that Helmholtz Theory was left to be forgotten, and the non- electromagnetic aspect of electricity was thereafter ignored by most with the exception of Nikola Tesla, and possibly Charles Steinmetz.

Heaviside concludes with:

"It is not by any means to be concluded that Maxwell spells finality. There is no finality. It cannot even be accurately said that Hertzian Waves prove Maxwell's Dielectric Theory completely. The observations are very rough indeed, when compared with the refined tests in other parts of electrical science. The important thing proved is that electromagnetic waves in the Aether at least approximately in accord with Maxwell's Theory are a reality, and that the Faraday-Maxwellian is the correct one." [25]

However, Heaviside imposes an arbitrary restriction:

"The other kind of electrodynamic speculation is played out completely. There will be plenty of room for more theoretical speculation, but it must be on the Maxwellian type to be really useful". [25]

Heaviside is correct in his criticism of the old German electrodynamics, the useless potential theories, and other such illusions in electrical theory, which find their origin as far back as Isaac Newton. However, to ignore the aspects of electrical theory, which are not of an electromagnetic nature is not permissible when framing a general theory of electricity. It is a strange irony that while Heaviside rejects the Helmholtz Theory, [33] his so-called Telegraph Equation does provide the required framework for a true General Theory of Electricity. For an excellent analysis of the development of Electric Transmission Theory, see Ernst Guillemin "Communications Networks", Volume Two, Chapter One.

The Third Commandment - Idolization Of The Electron

To Quote Lucien Poincaré:

"The electron has conquered physics and many adore the new idol rather blindly". [26]

Gustave Le Bon states:

"The conception of the electrons, a near relative to the old Phlogiston, is one of the most unfortunate metaphysical ideas recently formulated..." The electron has become at present day, a sort of fetish for many physicists, by which they think to explain all phenomena." [27], [34]

In a toast to a gathering of his illustrious colleagues, J. J. Thomson makes a rather startling remark:

"Here is to the electron, may it be of no use to anyone." [28]

It must be remembered that it was Thomson himself that is credited with the quantification of the so- called electron, so what would motivate him to make such an outlandish statement? [35]

Charles Steinmetz outright rejected the electron concept, or what he sometimes called the "Ionic Theory of Electricity." He regarded the notion of electronic charge as an impediment to the understanding of the behavior of electricity, to quote:

"Unfortunately, to a large extent in dealing with the dielectric field, the pre-historic conception of the electro-static charge on the conductor still exists, and by its use destroys the analogy between the two components of the electrical field, the magnetic and the dielectric and makes the consideration of the dielectric field unnecessarily complicated." [29]

After the introduction of the myth of Santa Clause, we as children are also "taught" that "electricity is the flow of electrons." It is later learned the truth of Santa, but the fallacy of the electron persists.

Oliver Heaviside writes extensively on the misguided notion that electricity is the flow of electrons in a so-called conductor. "A perfect conductor is a perfect obstructer, but does not absorb the energy of the electromagnetic wave…"

"The properties of a perfect conductor are derived from those common conductors by examining what would happen if the resistivity were continuously reduced, and ultimately became zero. In this way, we find that a perfect conductor is a perfect obstructer, for one thing, which idea is singularly at variance with popular notions regarding conductors…"

"According to Ohm's Law alone, a perfect conductor should be one which carried an infinite current under a finite voltage, and the current would flow all through it because it does so ordinarily. But what is left out of consideration here is the manner in which the assumed steady state is established. If we take this into account, we find that there is no steady state when the resistance is zero, for the variable period is infinitely prolonged, and Ohm's Law is out of it, so far as the usual application goes..."

"The smaller the resistance the greater the time taken for the current to get into the conductor from its boundary, where it is initiated. In the limit, with no resistance, it never gets in at all. Where then is the current?" [30]

Heaviside continues:

"The uniformly distributed current of the steady state appropriate to finite conductivity becomes a mere surface current when the conductivity is infinite."

"In the usual sense that an electrical current is a phenomenon of matter, it has become quite an abstraction, for there is no matter concerned in it. It is shut out completely." [31]

The electron represents a phenomenon of electronics, but not a phenomenon of electricity. Moreover, the motion of electrons is representative of the time rate of the destruction of electric induction; this is what is known as resistance. The concept of electrons finds, however, useful application in the theory of so-called semi-conductors, such as silicon, carbon, selenium, etc.

Where it is found in practice that electrical devices do not consume energy but serve to convert this energy into another useful form, electronic devices in practice consume nearly all the energy, turning it into useless heat.

The contemporary Leibnitz-Einstein notion of the electron has worked further mischief into electrical theory by it becoming a "catch all" for kindred phenomena, such as the cathode ray, etc.

The electron has become so endeared to the physicist that any criticism of it will draw an almost fanatical reaction.

What has been learned here is that in its fundamental nature, electricity is not a physical phenomenon married to gross physical matter, and such should be left aside in the Science of Electricity, and the ill-advised pronouncements of the physicist should be ignored.

REFERENCES

[1] Recent Researches In Electricity And Magnetism, J. J. Thomson, 1898, Page 2

[2] A History Of The Theories Of Aether And Electricity, E. T. Whittaker, 1910, Page 36

[3] The Evolution Of Force, Gustave Le Bon, 1908, Page 35

[4] Electromagnetic Theory, Oliver Heaviside, 1950, Article 179, Page 221

[5] Aether And Matter, J. Larmor, 1900, Page VIII

[6] Aether And Matter, J. Larmor, 1900, Page 23

[7] A History Of The Theories Of Aether And Electricity, E. T. Whittaker, 1910, Pages 1-3

[8] Electric Waves, Heinrich Hertz, 1893, Page XV

[9] Electromagnetic Theory, Oliver Heaviside, 1950, Page XIX

[10] Recent Researches In Electricity And Magnetism, J. J. Thomson, 1898, Page 1

[11] The Evolution Of Matter, Gustave Le Bon, 1906, Page 218

[12] The Evolution Of Matter, Gustave Le Bon, 1906, Page 198

[13] The Evolution Of Force, Gustave Le Bon, 1908, Page 109

[14] Electromagnetic Theory, Oliver Heaviside, 1950, Page XX

[15] Old Physics For New, T. E. Phipps Jr., 2012, Page 5

[16] Electric Waves, Heinrich Hertz, 1893, Page 15

[17] Theory Of Wireless Power, Eric P. Dollard, 1986, Page 1

[18] Charles P. Steinmetz, Engineer And Socialist

[19] A History Of The Theories Of Aether And Electricity, E. T. Whittaker, 1910, Page 309

[20] A History Of The Theories Of Aether And Electricity, E. T. Whittaker, 1910, Page 353
[21] Electric Waves, Heinrich Hertz, 1893, Page 20
[22] Electromagnetic Theory, Oliver Heaviside, 1950, Article 5, Page 5
[23] Electromagnetic Theory, Oliver Heaviside, 1950, Page XXVII
[24] Recent Researches In Electricity And Magnetism, J. J. Thomson, 1898, Pages 1-3
[25] Electromagnetic Theory, Oliver Heaviside, 1950, Article 6, Page 6
[26] The New Physics And Its Evolution, Lucien Poincaré, 1906, Page 324
[27] The Evolution Of Matter, Gustave Le Bon, 1906
[28] Proceedings Of The Royal Institution Of Great Britain, Vol 35, 1951, Page 251
[29] Electric Dischargers, Waves, And Impulses, C. P. Steinmetz, 1914, Pages 13-14
[30] Electromagnetic Theory, Oliver Heaviside, 1950, Article 189, Pages 338-339
[31] Electromagnetic Theory, Oliver Heaviside, 1950, Page 339
[32] The Evolution Of Matter, Gustave Le Bon, 1906
[33] Electromagnetic Theory, Oliver Heaviside, 1950, Appendix D, Page 493
[34] The Evolution Of Matter, Gustave Le Bon, 1906, Page 227
[35] Beyond The Electron, J. J. Thomson, 1928

LEARN MORE ABOUT THE LEGENDARY WORK OF

ERIC P. DOLLARD

Please support Eric Dollard & EPD Laboratories, Inc., a 501(c)3 non-profit corporation by making a tax-deductible donation at:

https://ericpdollard.com

For more books and videos by Eric Dollard, visit

https://emediapress.com/epd

Follow Eric on social media

https://ericpdollard.com/facebook

https://ericpdollard.com/twitter

Watch Eric's free videos and interviews

https://ericpdollard.com/youtube

Eric Dollard's Official Forum

https://ericpdollard.com/forum

http://www.energeticforum.com

https://www.energyscienceforum.com

EMEDIA
PRESS

www.ingramcontent.com/pod-product-compliance
Ingram Content Group UK Ltd.
Pitfield, Milton Keynes, MK11 3LW, UK
UKHW021926190726
13853UKWH00002B/864

9 798746 918185